All About Animals

Harcourt
SCHOOL PUBLISHERS

Orlando   Austin   New York   San Diego   Toronto   London

Visit *The Learning Site!*
www.harcourtschool.com

Animals Are Living Things

All animals are living things.
Rocks and water are nonliving things.

What Animals Need

Animals need food and water.
Most animals need shelter.

Air

gills

All animals need air.
Lungs help some animals breathe air.
Fish use gills to take in air.

Mammals

Mammals have hair or fur.
They give birth to live young.

Birds

Birds have feathers and wings.
Young birds hatch from eggs.

Reptiles and Amphibians

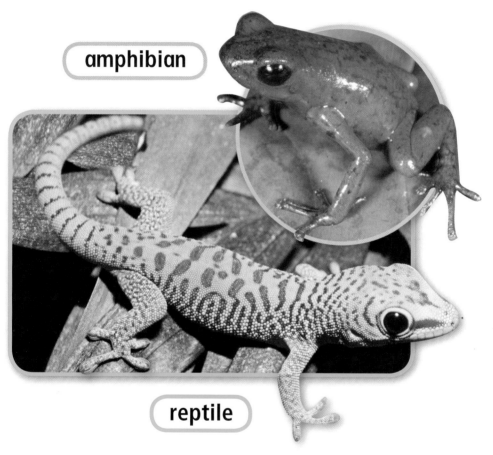

amphibian

reptile

Reptiles have scaly, dry skin.
Amphibians have smooth, wet skin.

Fish

Fish live in water.
Most fish have scales.

Insects

Insects have six legs.
They have three body parts.

Life Cycle of a Frog

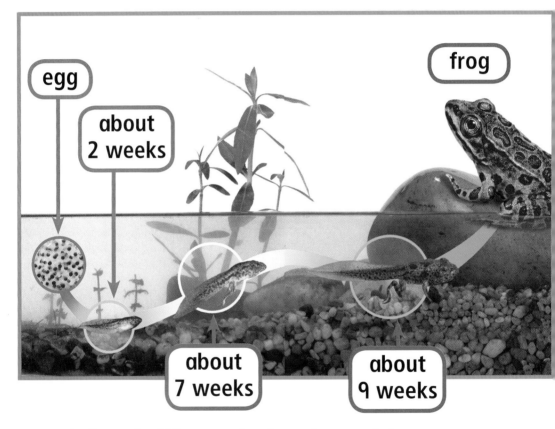

egg

frog

about 2 weeks

about 7 weeks

about 9 weeks

A frog's life cycle begins with an egg.
A tadpole grows from the egg.

Life Cycle of a Butterfly

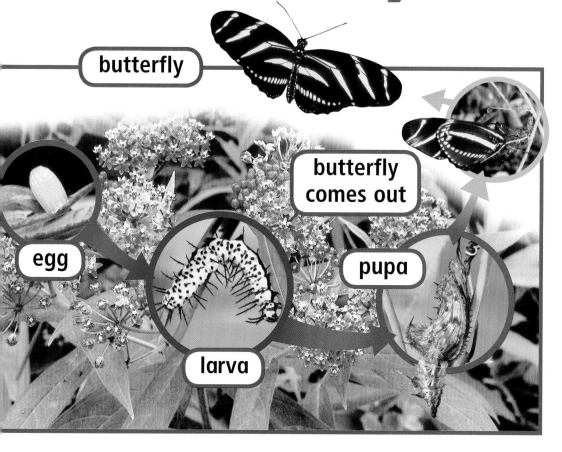

butterfly

butterfly comes out

egg

pupa

larva

A larva grows from an egg.
It becomes a pupa, then a butterfly.

Vocabulary